Reading Essentials in Science

CHEMISTRY CLUES

Mixtures and Solutions

JENNY KARPELENIA

PERFECTION LEARNING®

Editorial Director: Susan C. Thies
Editor: Mary L. Bush
Design Director: Randy Messer
Book Design: Emily J. Greazel
Cover Design: Michael A. Aspengren

A special thanks to the following for his scientific review of the book:
Kristin Mandsager, Instructor of Physics and Astronomy, North Iowa Area Community College

Image credits:
©AP: p. 14; ©Bettmann/CORBIS: p. 16 (bottom); ©DK Limited/CORBIS: p. 22; ©Reuters/CORBIS: p. 25; ©Stock Connection: p. 12; ©Holsten/Koops–StockFood Munich/Stockfood America: p. 20

BrandX Royalty-Free: pp. 7, 8 (top); Corel Professional Photos: p. 5 (bottom); ©Perfection Learning: back cover, pp. 3, 5 (top), 6, 8 (bottom), 10, 11, 13 (bottom), 18, 21, 23, 24, 26; Photodisc Royalty-Free: front cover, pp. 1, 13 (top), 17; Photos.com: pp. 9, 15, 16 (top), 27; Superstock Royalty-Free: p. 4

Text © 2006 by Perfection Learning® Corporation.
All rights reserved. No part of this book may be reproduced, stored in a retrieval system, or transmitted in any form or by any means, electronic, mechanical, photocopying, recording, or otherwise, without prior permission of the publisher.
Printed in the United States of America.

For information, contact
Perfection Learning® Corporation
1000 North Second Avenue, P.O. Box 500
Logan, Iowa 51546-0500.
Phone: 1-800-831-4190
Fax: 1-800-543-2745
perfectionlearning.com

1 2 3 4 5 6 PP 10 09 08 07 06 05

Paperback ISBN 0-7891-6618-6
Reinforced Library Binding ISBN 0-7569-4642-5

Contents

1. A SCIENTIFIC PICNIC 4
2. MIX IT UP 7
3. FINDING A SOLUTION 10
4. MORE ABOUT MIXTURES AND SOLUTIONS 15
5. SEPARATE IT 20

INTERNET CONNECTIONS AND
RELATED READING FOR MIXTURES AND SOLUTIONS . . 28

GLOSSARY 29

INDEX 32

CHAPTER 1

A Scientific Picnic

It's a warm, sunny day. The weather is perfect for a family picnic. When you arrive at the park, the first thing you do is check out the food. Several picnic tables are covered with favorite family dishes. Your mom made fruit salad. You had to help cut up the watermelon and cantaloupe. Then you mixed it with grapes, blueberries, and strawberries. Your grandma brought some of her famous potato salad. It has big chunks of potatoes, slices of hard-boiled eggs, bits of bacon, and pieces of celery and onion. You grab a handful of a snack mix. It has all of your favorites—peanuts, popcorn, pretzels, raisins, chocolate chips, and M&Ms. The sweet and salty mix makes you thirsty. To quench your thirst, you head to the drinks. You have your choice of water, Kool-Aid, or soda.

THE SCIENCE IN YOUR PICNIC

What does science have to do with your family picnic? Many of the foods and drinks at your picnic are **mixtures**. A mixture is a combination of two or more substances. The substances in a mixture keep their own chemical properties. They act the same as if they were separate. When you eat a peanut out of the snack mix, it's no different than if you ate one straight out of the shell.

The "ingredients" in a mixture are not chemically bonded together. They don't undergo any chemical reaction to make a new substance with different properties. To make the hard-boiled egg for the potato salad, you had to cook a raw egg. The heat caused a chemical reaction in the egg and turned it into a new substance. But the boiled egg in the potato salad didn't change when it was added to the potato salad. It was a hard-boiled egg when you sliced it for the salad, and it was still a hard-boiled egg once it was mixed in with the potatoes and other ingredients.

The materials in a mixture can be mixed together in different amounts. One fruit salad might have a cup of each kind of fruit. Another might have two cups of some fruits and half a cup of others.

Mixtures can be **heterogeneous** or **homogeneous**. The fruit salad, potato salad, and snack mix are examples of heterogeneous mixtures. *Heterogeneous* means "different throughout." A heterogeneous mixture has parts that aren't evenly mixed. Every bite of fruit salad will have different amounts of watermelon, grapes, and strawberries. Every handful of snack mix will have a different number of pretzels, chocolate chips, and M&Ms.

The drinks at your picnic are examples of homogeneous mixtures. *Homogeneous* means "the same throughout." The materials in a homogeneous mixture are spread out evenly in the mixture. One drink of soda is the same as the next.

Homogeneous mixtures are also known as **solutions**. Solutions are simply mixtures that are evenly mixed. Most solutions are formed from dissolving a solid, liquid, or gas in a liquid. The Kool-Aid flavoring and coloring powder dissolves in water. The sugar that is already in the Kool-Aid or that you add separately also dissolves in the water. All of the parts are mixed together evenly. You don't get a drink of flavoring in one spot and a drink of sugar in another. That's a perfect scientific solution for your thirst!

CHAPTER 2

Mix It Up

Heterogeneous mixtures surround you everywhere. Many of the foods you eat are mixtures. Some of the clothes you wear are made of mixtures. Rocks and sand are mixtures. Even the soil and concrete you walk on are mixtures. Try getting through the day without mixtures and you won't get far!

FOODS AND FABRICS

Some of your favorite foods may be mixtures. Trail mixes, cereals, and pasta salads are mixtures. A meal of tacos, lettuce salad, and rocky road ice cream consists entirely of mixtures. The tacos are a mixture of taco shells, hamburger, cheese, lettuce, and other toppings. Lettuce, carrots, tomatoes, croutons, and other toppings are mixed together to make the salad. A mixture of chocolate ice cream, marshmallows, and nuts creates a tasty treat for dessert. How many meals can you think of that are made of mixtures?

Warning! Cooking Causes Chemical Changes

Be careful not to confuse foods that are a result of chemical reactions and those that are mixtures. Foods that are cooked have undergone a chemical change. These are not mixtures. However, mixtures can be made of foods that were previously cooked. For example, the hamburger in your tacos was previously cooked, or chemically changed. However, it isn't chemically changed when it's mixed with the other taco ingredients. You can separate the cooked hamburger from the rest of the taco. This makes the taco a mixture.

Take a look at the tags on some of your clothes. You'll notice that many are a mixture of fabrics. Many clothing articles are cotton/poly blends. This means that they are a mixture of natural cotton threads and artificial polyester threads woven together. Running shorts are often a mixture of cotton and spandex or nylon and spandex.

ROCKS AND SAND

Rocks are mixtures of different minerals. Sandstone is a mixture of sand grains. Granite is a mixture of quartz, feldspar, and mica. Shale is a mixture of mud and clay. Conglomerate and breccia are mixtures of different-sized rock fragments.

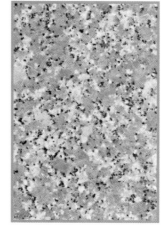
Granite

Conglomerate

Sand is a mixture of loose rock and mineral grains. If you examine a handful of sand, you'll spot many different-sized grains of various colors.

SOIL AND CONCRETE

Take a look under your feet and you'll more than likely find a mixture. Soil is a mixture of small rock and mineral pieces. It also contains bacteria and decaying plant and animal material. Concrete is a mixture of **cement**, sand, water, and ground-up rocks and minerals.

CHAPTER 3

Finding a Solution

Homogeneous mixtures are called *solutions*. The salt water in the ocean and the air in the atmosphere are solutions. Solutions are created when one substance is **dissolved** in another substance. The substance that is dissolved is the **solute**. The substance that is doing the dissolving is the **solvent**.

Water is often used as a solvent. In fact, water is known as the "universal solvent" because it can dissolve more substances than any other material. If you stir a tablespoon of sugar into a glass of water, you make a sugar-water solution. The sugar is the solute. The water is the solvent.

In a solution, each solute particle is surrounded by many solvent particles. When you stir sugar into water, for example, the water **molecules** move around to make room for the sugar molecules. As the molecules interact, they eventually settle into an evenly mixed arrangement.

Solids, liquids, and gases can all be solutes and solvents in a solution. Salt water is a solid/liquid solution (salt in water). Vinegar is a liquid/liquid solution (acetic acid in water). Soda is a gas/liquid solution (carbon dioxide in water). Air is a gas/gas solution (oxygen and several other gases in nitrogen gas). Steel is a solid/solid solution (carbon in iron).

THE STRENGTH OF THE SOLUTION

In most solutions, there is more solvent than solute. However, the exact amounts of solvent and solute vary with each solution. When there is very little solute dissolved in a large amount of solvent, the solution is said to be **dilute**. As the amount of solute increases in a given amount of solvent, the solution becomes more **concentrated**.

A cup of tea is a good example of dilute and concentrated solutions. If you place a tea bag in a cup of hot water for 30 seconds, only some of the tea will dissolve in the water. You will have a dilute tea solution. If you leave the tea bag in the water longer, more of the tea will dissolve. The longer you leave the tea bag in the water, the more concentrated the tea becomes. It will look darker in color and have a stronger taste.

HOW MUCH CAN IT HOLD?

At a given temperature and pressure, every solvent can only hold a certain amount of solute. An **unsaturated** solution does not contain the maximum amount of solute possible. More solute can still be dissolved in the solvent. When the solvent can no longer dissolve any more solute, the solution is **saturated**. If more solute is added after a solution is saturated, the solute will not dissolve. It will begin to settle out instead.

Most solid/liquid solutions can hold more solute when the temperature is increased. For example, if you heat up a saturated sugar-water solution, you can dissolve more sugar in the water.

The opposite is true with gases. As you increase the temperature of a gas solution, it can hold less solute. However, an increase in pressure will increase the amount of solute that can be dissolved. Soda is a good example of this. When contained in the can, the pressure is great, so a larger amount of carbon dioxide gas can be dissolved in the water. However, if you open the can and decrease the pressure, the amount of solute that the water can hold decreases. The excess gas is released into the air. That's why there's a sudden rush of "fizz" when you pop open a soda can.

A Super Trick

There is a way to "trick" a solid/liquid solution into holding more solute. If the solution is heated, more solute will dissolve. When cooled to the original temperature, the solution will hold the additional solute. The solution is now **supersaturated**.

SPEEDY SOLUTIONS

The size of the solute affects how fast it dissolves in a solvent. In general, larger solutes take longer to dissolve than smaller ones. Try dissolving a sugar cube in a glass of water. Time how long it takes for the sugar to dissolve. Then time how long it takes a crushed sugar cube to dissolve in the same amount of water (at the same temperature). You'll see that the crushed sugar cube dissolves faster than the whole one. That's because the water is able to surround more sugar molecules more quickly when they are separate, smaller pieces. In a whole cube, the outer molecules must be dissolved first before the inner ones even have contact with the water.

Stirring a solution also increases dissolve time. The motion creates more opportunity for the solvent molecules to surround the solute.

Temperature and motion are both important factors for gases dissolved in liquids. As a gas gets hotter and more agitated (stirred up), less of it can be dissolved. The same is true of a liquid. If it is agitated, it will hold less gas. This is because increased heat and motion cause gas molecules to escape into the air instead of being dissolved in the liquid.

HOTTER OR COLDER

Adding a solute to a solvent affects the boiling and freezing points of the solvent. The addition of a solute increases a solvent's boiling point. For example, plain water boils at 212°F. When salt is added to the water, it takes a higher temperature to boil the solution. The amount of salt added to the water affects the temperature increase. Greater amounts of salt result in higher boiling points.

A solute has the opposite effect on a solvent's freezing point. Adding a solute lowers the freezing point. While plain water freezes at 32°F, salt water freezes at colder temperatures. Again, the actual temperature decrease depends on the concentration of salt in the water. The more concentrated the solution is, the lower the freezing point is.

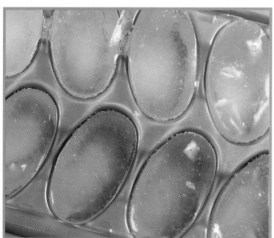

13

Have you ever wondered why salt is sprinkled on icy sidewalks? Besides increasing your traction, the addition of salt changes the freezing point of the water. When some of the ice melts, it forms a saltwater solution that will have to get even colder to refreeze.

A Freezing Solution

The antifreeze in a car's radiator is a solution of propylene (or ethylene) glycol, water, and a coloring agent. Antifreeze helps keep the water in the radiator and engine block from freezing in cold temperatures because the solute, propylene glycol, lowers the freezing point. It now has to get even colder than 32°F for the solution to freeze and crack the radiator or engine block.

CHAPTER 4

More About Mixtures and Solutions

Several types of mixtures have special properties and names. They are **suspensions**, **colloids**, **emulsions**, and **alloys**.

SUSPENSIONS

When the particles in a mixture are large enough to see with your eyes and settle out after standing, the mixture is called a *suspension*. Suspensions usually appear **opaque**, murky, or hazy. This is because light traveling through them is scattered by the suspended particles.

Dirt can form a suspension in a water puddle. The dirt will hang in the water, making it look "muddy." But if the puddle is left undisturbed, the dirt will eventually settle to the bottom.

Dusty air is also a suspension. When you dust your house, you stir up dust and dirt particles. They mix with the air. You can usually see them hanging there immediately after you sweep them off your furniture. But after a while, those dust particles will settle out of the air and land on your stuff again. That's why dusting is an endless chore!

Have you ever noticed the tiny particles hanging in the air around a movie projector's beam of light? These are also dust particles suspended in the air. The particles scatter the light from the projector as it makes its way to the screen. This scattering of light by particles in its path is known as the Tyndall Effect.

Scientist of Significance

John Tyndall (1820–1893) was a scientist, teacher, and inventor from Ireland. He studied both chemistry and physics and made many significant discoveries in both areas. It was Tyndall who first recognized that light was only visible when it bounced off particles in the air. His experiments showed that different-sized particles caused light to scatter in different ways. Tyndall also experimented with water vapor, carbon dioxide, and the ozone layer to find out how these substances absorbed the Sun's heat.

Tyndall's inventions include a breathing apparatus for firefighters and a light pipe. The light pipe was the forerunner of today's fiber optics technology.

COLLOIDS

A colloid is a mixture that has slightly smaller particles than those of a suspension. Some colloids look cloudy or murky. Others may appear evenly mixed to the human eye. However, the particles are actually not dissolved in the mixture. Instead, they are very small and keep colliding with one another. This continual bumping keeps the colloid mixed.

The particles in a colloid are too small to be **filtered** out of the mixture. They also do not settle out on their own. They are still large enough, however, to scatter light.

The blood running through your body is a colloid. Tiny red and white blood cells are suspended in liquid plasma. The cells do not settle out of the blood.

Fog is another example of a colloid. Fog is clumps of water molecules mixed in air. The water molecules scatter light that shines through the fog. That's why the light from a car's headlights is distorted by fog.

17

EMULSIONS

An emulsion is a special type of colloid made from tiny liquid droplets suspended in another liquid. These droplets will scatter light just like the particles in other suspensions and colloids. Over time, the liquids in an emulsion will separate.

Salad dressings made with oil and vinegar are emulsions. When sitting on a store shelf, the oil and vinegar are in separate layers. When you shake the bottle, the oil and vinegar mix to form an emulsion. When the bottle of dressing sits in the refrigerator for a while, the oil and vinegar will separate again.

The milk you buy from the grocery store is also an emulsion. Tiny particles of protein and butterfat are mixed in the watery milk that comes fresh from the cow. If left alone, milk will separate into a layer of skim milk and a layer of fatty cream. To make this separation take longer, fresh milk undergoes a process called *homogenization*, or extreme mixing. This keeps the emulsion mixed for a much longer time.

> ## Inquire and Investigate: The Tyndall Effect in Mixtures
>
> **Question:** How does light travel through mixtures?
> **Answer the question:** I think that light _____.
> **Form a hypothesis:** Light _____.
> **Test the hypothesis:**
>
> Materials
> - 1 cup of tap water mixed with 1 teaspoon of salt
> - 1 cup of milk
> - strong, thin beam of light (laser pointer or powerful pen flashlight)
> - dark room
>
> Procedure
> ✸ Shine the light through each mixture and observe what happens.
>
> **Observations:** The salt water should let the light pass right through. You will not see the beam of light in the water because there are no particles to scatter it. You will see the beam move through the milk because the tiny particles suspended in the milk will scatter the light.
>
> **Conclusions:** Light travels straight through solutions but is scattered by the particles in other mixtures (suspensions, colloids, and emulsions). This scattering of the light makes it visible.

ALLOYS

An alloy is a solid/solid solution of a metal and at least one other **element**. Sometimes a pure metal has weaknesses. Pure gold is very soft. Pure iron will rust. But when mixed with different elements, these metals can be improved. Silver and copper are often mixed with gold to form a strong alloy used to make jewelry. Carbon is mixed with iron to form the strong alloy steel that is used in construction. Everyday silverware is usually stainless steel, an alloy of iron, carbon, chromium, and nickel. The brass instruments in a band are alloys of copper and zinc. Many coins are now alloys of copper, nickel, and zinc to make them more durable. Even the fillings in your teeth may be alloys of mercury, silver, zinc, and tin. These alloys made with mercury are called *amalgams*.

CHAPTER 5

Separate It

Because mixtures are not chemically combined, they can be separated into their individual parts. The fruit salad at your picnic, for instance, could be separated just by picking out each of the fruit pieces. If you let the Kool-Aid sit out in the Sun long enough, the water would evaporate, leaving the coloring, flavoring, and sugar behind.

In general, the difficulty of the separation process depends on the size of the particles. Mixtures with larger particles are generally easier to separate. For example, it would be easier to separate your picnic snack mix than the soda.

Different methods are used to separate different types of mixtures. Some of these methods involve sorting particles by size. Others separate mixtures by using certain properties of the individual materials. Several methods use heat to separate substances.

SIEVING

Sieving is a common method used to separate mixtures with larger solid particles. A sieve (also known as a sifter or strainer) is a device with openings of a certain size. When a mixture is poured through it, a sieve will only allow particles that

are smaller than the openings to pass through. Larger particles remain on top. Different-sized particles can be sorted out of a mixture by using sieves with progressively smaller openings. This method works well when separating a mixture of rocks, gravel, and sand.

FILTERING

Filtering is a way to separate mixtures of solids mixed in liquids or gases. A **filter** is a material such as sand, paper, or cloth that has pores (tiny holes) in it. The pores are big enough to allow the liquid or gas to pass through but are too small for larger solid particles to fit through. A coffee machine works by straining water through a filter holding coffee grounds. The filter allows the hot water to pass through and into the coffeepot. The hot water dissolves some of the flavoring and coloring from the coffee grounds on its way through the filter. The coffee grounds themselves are too big to fit through the filter, so they stay out of the pot.

The furnace used to heat your home most likely contains an air filter. The filter is used to capture solid particles (dust, pet hair, smoke) from the air in your home. This purifies the air so that it's healthier to breathe as it's recycled again and again.

CHROMATOGRAPHY

Chromatography is another separation technique. *Chromatography* means "color writing." Paper chromatography separates the different-colored **pigments** that make up a solution by measuring and comparing how fast the colors move through chromatography paper.

Chromatography is also used to analyze other mixtures (air, perfume, blood, urine, etc.) to find out what chemicals are in them. A mixture is passed through a gel (instead of paper), and different chemicals in the mixture (pollutants in the air, proteins in blood, or drugs in urine) are separated out as they move through the gel at different speeds.

Inquire and Investigate: Chromatography

Question: What colors are mixed together to make black?
Answer the question: I think that black is a mixture of _____.
Form a hypothesis: Black is a mixture of _____.
Test the hypothesis:

Materials
- black water-soluble (not permanent) marker
- coffee filter
- plate
- water
- eyedropper

Procedure
✻ Draw a black dot about the size of a dime in the center of the coffee filter. Place the filter on the plate. Squirt a few drops of water on the dot and observe for a few minutes.

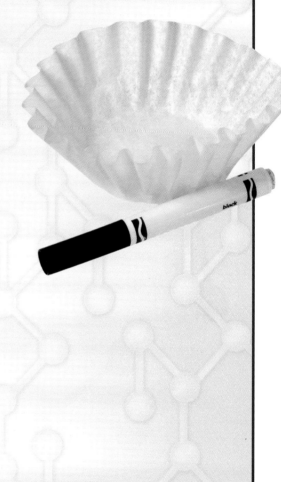

Observations: The water and ink move upward through the filter paper. As they move, the black ink is separated into green, blue, red, and yellow. Each color is carried by the water at a different speed depending on the color's molecule size and how strongly it is attracted to the filter paper.

Conclusions: Black is a mixture of green, blue, red, and yellow.

DECANTING

Decanting is a procedure used to separate suspensions. In a suspension, the particles eventually settle out of the mixture. When this happens, the liquid can be slowly and carefully poured off the top. This separates the solid particles from most of the liquid. Imagine you're on a camping trip and the only water around is from a muddy river. You could clean the water by decanting it (if you have the patience). After you wait for the rocks, sand, and soil to settle at the bottom of the glass, the water left on top would be much cleaner to drink. Wine is decanted to separate the liquid from any particles that may settle out in the bottle. This makes the wine taste better.

Decanting can also be used to separate two liquids in a suspension. For example, an oil and water mixture can be separated by slowly pouring the oil layer off the water.

Technology Link

A centrifuge (SEN truh fyouj) is a tool used to separate suspensions, colloids, and emulsions quickly. A centrifuge spins the mixture at high speeds. The spinning force pulls the denser particles to the bottom. The lighter particles stay on top. Blood is often centrifuged to separate the blood cells from the plasma. Fresh milk is also centrifuged to separate the watery part from the fatty part. Then the amount of fat added back to the milk can be controlled to create skim, low-fat, or whole milk.

MAGNETISM

If some parts of a mixture are magnetic and others are not, a magnet can be used to pull the magnetic parts out of the mixture. This separation method is used to pull recyclable iron and steel materials out of a pile of garbage. You can actually use magnets to separate iron out of some foods. Try crushing several cups of an iron-enriched cereal into a powder. Pour the cereal into a large bowl and add hot water to make a "soup." Stir the cereal mixture with a white bar magnet for several minutes. You should notice fuzzy black pieces of iron sticking to the magnet.

HEAT

Heat can also be used to separate mixtures, especially when the particles are too small to be removed by another method. Heat causes liquids to **evaporate**, leaving the solid particles behind. Salt water can be separated this way. Mix a teaspoon of salt into a small bowl of water and stir. Let the solution sit in a sunny window for a few days. You should see the salt left behind in the bowl as the water gradually evaporates into the air.

Boiling can speed up the separation process. **Distillation** is the separation of a solution by boiling and cooling. It is used with mixtures that are made of materials with different boiling points.

Distilled water is purified this way. Water is boiled so evaporation occurs quickly. Minerals and other impurities are left behind and removed. The evaporated water is then **condensed** back into a liquid. This process can be used on a large scale to take the salt out of ocean water so it's drinkable.

An Example of Drinking Water Distillation

1. Ordinary tap water is heated to 212°F, killing bacteria and viruses that may be present.
2. Steam rises, leaving behind dissolved solids, salts, heavy metals, and other substances.
3. The steam is condensed in a stainless steel coil.
4. Liquids with lower boiling points than water stay a vapor and are released through a vent.
5. The distilled water flows through a filter that removes any remaining impurities.
6. The purified drinking water is collected in a container.

Distillation can also be used to separate liquids with different boiling points. When a mixture is boiled, each liquid will evaporate at a different temperature. One by one, each gas is then condensed and collected. Distillation is used to separate the thick black crude oil that's pumped from deep within the Earth. Crude oil is a mixture of different liquids such as gasoline, diesel fuel, propane, and kerosene. Each of the separated liquids can then be used as a fuel or to make products such as plastics and waxes. The heaviest part of crude oil can be used to make asphalt to surface roadways.

Tracing the Path of Science

Before the mid-1800s, light came from torches, candles, and lamps that burned whale oil. In the 1840s, Albert Gesner changed all that. He was the first person to distill kerosene from crude oil (also known as petroleum). Quickly, kerosene lamps replaced other sources of light. Kerosene remained in heavy demand until the lightbulb was invented in 1879. Gesner's distillation technique led to the creation of other petroleum products. Large quantities of gasoline, in particular, became a necessity in the early 1900s when cars were invented. Gesner soon became known as the "father of the petroleum industry."

Mix them together. Take them apart. Put them back together again. Mixtures and solutions are very versatile combinations that you depend on every day.

Internet Connections and Related Reading for Mixtures and Solutions

http://www.chemkids.com/files/matter_intro.htm
Explore mixtures, alloys, and solutions at this Chem4Kids site.

http://www.elmhurst.edu/~chm/vchembook/106Amixture.html
Simple experiments demonstrate the basic principles of mixtures and solutions.

http://www.infoplease.com/ce6/sci/A0812901.html
Get caught up on colloids with this basic information.

http://www.bbc.co.uk/schools/revisewise/science/materials/10_fact.shtml
Find out the facts about separating materials, and then complete activities and a quiz to see how much you've learned.

Elements, Compounds and Mixtures by J. M. Patten. A book about what elements, compounds, and mixtures are and how they are useful in everyday life. Rourke Book Company, Inc., 1995. [RL 3 IL 1–3] (0218606 HB)

Matter by Christopher Cooper. This Eyewitness Science Book on matter includes discussions on mixtures, alloys, and colloids. Dorling Kindersley, 1992. [RL 8.2 IL 3–8] (5869206 HB)

- RL = Reading Level
- IL = Interest Level

Perfection Learning's catalog numbers are included for your ordering convenience. HB indicates hardback.

Glossary

alloy (AL oy) solution of a metal and at least one other element (see separate entries for *solution* and *element*)

cement (si MENT) limestone rock and clay ground into a fine powder

chromatography (krohm uh TAHG ruh fee) process used to separate a mixture by passing it through or over something that absorbs the different parts at different rates

colloid (KAHL oyd) mixture with small particles suspended in another substance

concentrated (KAHN sen tray ted) containing a large amount of solute in proportion to solvent (see separate entries for *solute* and *solvent*)

condense (kuhn DENS) to change from a gas to a liquid

decanting (dee KANT ing) method that separates a mixture by pouring the liquid off

dilute (deye LOOT) containing a small amount of solute mixed in a large amount of solvent (see separate entries for *solute* and *solvent*)

dissolve (diz AWLV) to evenly mix with another substance

distillation (dis til LAY shuhn) process of separating a mixture by boiling off the liquid(s)

Glossary

element (EL uh ment) nonliving material made up of one type of particle

emulsion (im UHL shuhn) mixture of tiny liquid droplets suspended in another liquid

evaporate (ee VAP or ayt) to change from a liquid to a gas

filter (FIL ter) to put a mixture through a filter to separate its parts (verb); device for separating larger particles out of a mixture (noun)

heterogeneous (het er oh JEE nee uhs) consisting of parts that are not evenly mixed

homogeneous (hoh moh JEE nee uhs) consisting of parts that are evenly mixed

mixture (MIKS cher) combination of two or more substances that keep their own properties when combined

molecule (MAHL uh kyoul) smallest unit of a substance that can exist by itself

opaque (oh PAYK) not allowing light to pass through

pigment (PIG ment) substance added to something to give it color

saturated (SATCH er ay ted) containing the maximum amount of dissolved solute possible (see separate entry for *solute*)

solute (SAHL yout) substance that is dissolved in another substance

solution (suh LOO shuhn) mixture with one or more substances dissolved uniformly in another substance

solvent (SAWL vent) substance that dissolves another substance

supersaturated (soo per SATCH er ay ted) containing more solute than normal at a certain temperature (see separate entry for *solute*)

suspension (suh SPEN shuhn) mixture with larger particles that settle out

unsaturated (uhn SATCH er ay ted) not containing the maximum amount of dissolved solute possible; able to dissolve more solute (see separate entry for *solute*)

Index

alloys, 19
colloids, 17
emulsions, 18
Gesner, Albert, 27
mixtures
    definition, 4–5
    heterogeneous, 6, 7–9
    homogeneous, 6, 10–14
separating mixtures, 20–27
    centrifuge, 25
    chromatography, 22, 23
    decanting, 24
    distillation, 26–27
    evaporation, 26
    filtering, 21
    magnetism, 25
    sieving, 20–21
solutions, 10–14
    boiling points, 13
    concentrated, 11
    dilute, 11
    dissolving rate, 12–13
    freezing points, 13–14
    saturated, 11, 12
    solute, 10
    solvent, 10
    supersaturated, 12
    unsaturated, 11
suspensions, 15–16
Tyndall Effect, 16, 19
Tyndall, John, 16